Wojtek Palubicki

Fuzzy Plant Modeling with OpenGL

Wojtek Palubicki

Fuzzy Plant Modeling with OpenGL

Novel Approaches in Simulating Phototropism and Environmental Conditions

VDM Verlag Dr. Müller

ISBN: 978-3-8364-2747-0

Table of Contents

1. Introduction

Plants have been part of this world for hundreds of millions of years, changing and adapting like all other life forms into the shapes and forms that we are now granted to enjoy. From the very beginning of human culture people developed an interest in these indispensable entities of the ecological cycle. Exploiting, naming and analyzing them countless of times in a multitude of various ways. Our curiosity never seemed to have stopped, as we are still uncovering their secrets with no end in sight.   Recently, with the development of computer-aided tools scientists are bound to widen the borders of our understanding even further. Complex data can be computed in a matter of moments while sophisticated rendering programs are able to deliver pleasing visual presentations of the results. From crop growth to forest development over to virtual landscape modeling, nowadays this information finds application in all manner of different fields. Such projects require the expertise of researchers with backgrounds in e.g. agriculture, forestry, physics, computer science or mathematics, making them highly interdisciplinary undertakings of considerable difficulty. The particular attempts often vary greatly in quality or are flawed by a one sided viewpoint, thus falling short to convince unanimously. The successful ones however are widely used with great regard in respect to the improvements they have on our living conditions.

Despite these considerations it is the ambitious goal of this work to present an attempt at modeling the growth of plants on modern computers with the aid of current 3D graphics possibilities. As part of it, a computer program has been developed which served both as a tool to deepen the understanding for the matter as well as means to accomplish necessary experiments along with it. All algorithms discussed here were implemented and tested, their results are presented throughout the pages of this work.

A simulation based approach dictated the course of the project. i.e. that not only visual results but also the functional modeling of underlying biological processes responsible for plant development were major goals. Additionally the growth of plants can be observed in a continuous manner with a selection of development strategies and environmental conditions available. Due to the immense nature of the problem and the diverse knowledge needed, only the functionally most important factors in plant development could be simulated without going out of resources. The proposed solutions are in many cases not as detailed and refined as earlier attempts that focused on a very narrow range of problems only. However, they represent novel ideas that while not solving particular issues associated with the field of research in a better way, produce visually viable results in real-time. Whereas many of the

2

technically advanced methods are restricted to rendering the plant models beforehand.

Nowadays there exist a wide array of approaches to model the growth of a plant, but fundamentally they all follow one of two different methodologies. One being the Structural model, treating the plant as a thing described purely by physical quantities, the other the Functional-Structural model, where the plant is understood as part of a process driven environment. Fractal mathematics invented by Mandelbrot, are a good example of the former while simulations under approximated real conditions for the latter. Functional-Structural approaches produce more convincing results, because a much larger information base is available. It is them who have been extensively expanded in the most recent years. Structural approaches make use of the self-similar forms of plants to generate them. Situations where a plant would break the rules of self-similarity due to the application of certain conditions (e.g. lack of light in one area of growth) are therefore out of scope for these methods.

In 1968 Lindenmayer introduced a parallel string rewriting system later coined L-Systems. It was conceived for biologists and mathematicians, but in 1988[5] extended for the use in computer graphics.

"The essence of development at the modular level can be regarded as a sequence of events, in which *predecessor* or *parent* modules are replaced by configurations of *successor* or *child* modules. The rules of replacement are called *productions*. It is assumed that the number of different module types is finite, and all modules of the same type behave in the same manner. Consequently, the development of a large structure (configuration of modules) can be characterized by a finite set of rules. An L-system is simply a specification of the set of all module types that can be found in a given organism, the set of productions that apply to these modules, and the initial configuration of modules (the *axiom*), from which the development begins."
[1] *P. Prusinkiewicz, leading researcher in L-Systems on L-Systems*

The formal grammar G = {V, ω, P} compromised of the triplet V (Vocabulary), P (Production rules) and ω (axiom) describes our L-System. Where V is our alphabet of characters containing replaceable elements, P a set of rewriting rules which turn a predecessor string into a successor string and ω the initial state of the system. Numerous extensions have been introduced throughout the course of years, most notably among them are context-sensitive L-Systems, non-deterministic/stochastic L-Systems, parametric L-Systems[10], Open L-Systems[3] and the so called dL-Systems (differential) [8] .

Context-sensitive L-Systems use production rules dependent on its neighbors in contrast of the original approach which was context free. Stochastic L-Systems have more than one production rule that applies to a current symbol or string, otherwise they are termed deterministic. Parametric L-Systems use parameters as an extension to symbols of its alphabet. dL-Systems define differential equations, solved using a tridiagonal coefficient matrix[9], that turn the discrete approach into a continuous one. Open L-Systems evolved from parametric and environmentally-sensitive L-Systems[12]. They sport a separate process, which simulates the environment and is capable of interchanging data with the plant growing process. Recent works include the introduction of interactively arranged L-Systems[13], where the plant is composed of spring connected branch parts, that are endowed with physical quantities. According to the authors a biologically faithful manipulation during growth process is therefore possible.

L-Systems propose the most extensively researched approach so far, presenting biologically realistic and aesthetic results. Plants[14], bacteria[9], sea shells[15] and many other life forms have successfully been modeled by researchers with various backgrounds. In addition, a number of libraries and tools, like cpfg and GROGRA, are available, further increasing its popularity. Despite these convincing arguments L-Systems suffer from a severe drawback. They are complex. Thus a lot of experience and time is needed to create the desired models. Furthermore small errors in defining the production rules can have strange effects on the overall outcome of the model, making debugging a painstaking undertaking. To remedy this situation many different generic models have been proposed, a brief summary of some well known growth models is presented below.

De Reffye *et al.* proposed in 1988 an approach based "on a botanically accurate simulation of the functioning of meristems"[6]. The procedural model is constructed around the growth, ramification and mortality of meristems. It was later expanded and improved and finally turned into the commercial tool AMAP. An established program, computing photo realistic plants in a biologically accurate way. The experiments presented later on in this thesis are leaned upon the same idea.

Another generic approach which was limited to trees only was introduced in 1994 by Weber and Penn[16]. This time botanical correctness was less of an issue, instead emphasis was laid on easily operable parameters, like shape of the tree or starting height of the branching process. That way, in a very short time span, users were able to generate remarkably realistic trees without prior botanical or programming experience. Further tools, like Arbaro, Treal and TreeDesigner were developed based upon the idea of the original work proving the success of the approach.

In the same spirit Deussen and Lintermann began work on a tool named xfrog, which was presented to the public in 1997[17]. With the genuine idea of using graphical components as edges of a user constructed graph, the program allowed to define key growth patterns and plant parts necessary to generate the plant. The graphical icons could describe algorithmic

4

operations as well as geometric primitives or material components. Only a handful of such components connected to a graph could already generate believable models.

Drawing from a wide array of algorithms Benes *et al.* proposed in 2002 a fine growth model dealing with climbing plants[2]. They used a botanically faithful approach incorporating notions from particle systems, which turned their plants into intelligent competitors for light and space. While the ideas to utilize ray casting to simulate phototropism[18], or the use of voxels for collision detection[19] weren't new, the combination of these algorithms gave extremely neat results. In many ways their work serves as an example of competent engineering, and is an ample source of inspiration and motivation for this work.

A totally different technique to render plants has been introduced in 1985 by Reeves and Blau[4]. They used particle systems coupled with stochastic processes to generate stunning forest scenes for the movie 'The Adventures of André and Wally B.'. Due to the considerable effort necessary in implementation and rendering, the technique never became a popular choice to model plants.

Overall there is a definite trend towards ever more complex simulations in the past years. The results are more complete and live up to a very high standard. Current research also includes the modeling of whole plant populations[7] and increasingly detailed environment simulations[37]. Although, despite all these efforts there are still a number of problems, which are less thoroughly researched. The generation of believable bark or the incorporation of a meteorological correct climate model are yet to be convincingly proposed.

Concerning the structure of this work, the author included a short introduction describing the basic botanical notions necessary for understanding, in the chapter immediately following this one; after which thorough discussion of the simulation itself begins. In which, first the key features of the model are presented, then follows a description of the environment model, afterwards some thoughts are given how to expand the approach towards whole populations of plants. Finally light is shed into the technical realization of the computer application.

2. Plant growth

Before delving any deeper into the matter a brief definition and explanation of terms associated with our problem is presented.

Plant growth is the result of the production of cellular tissue by an organ called the **merisstem** located in the inner layers of the **bud**. For the purposes of this thesis it is sufficient to assume that the bud is the main production organ. We distinguish the **apical bud** at the tip of a branch compared to the **axilar/lateral buds** located along the axis. Other typical plant organs include **leaves, fruits, flowers, thorns, tendrils, nodes** and **internodes**, of which all can be procreated by the bud. In the case of flowers, thorns, tendrils and fruits the bud dies in the process. Nodes are situated between two internodes and bear leaves and buds, whereas internodes are the parts that form the axis.

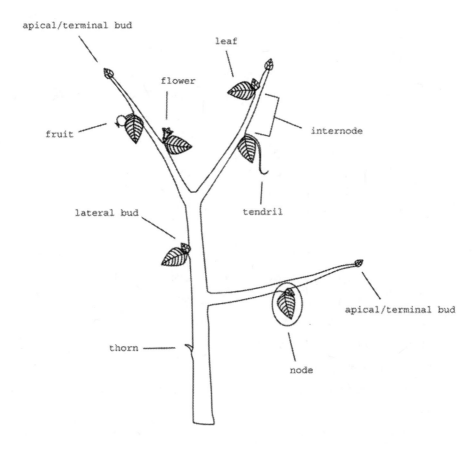

6

Phyllotaxis is a Greek term meaning "order of leaves", and refers to the regular laws that apply to the arrangement of leaves along the axis of a plant. The cause of this arrangement pattern is the rotation of the apical bud about it's own axis. Basic patterns include among many others: **distichous, multijugate, whorled** or **spiral**, depicted below.

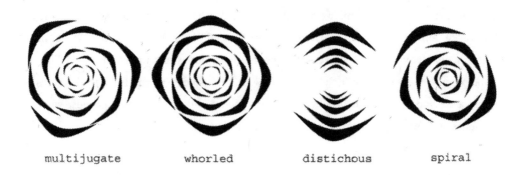

multijugate   whorled   distichous   spiral

The **stem development trend** is divided into **orthotropy** and **plagiotropy** as illustrated on the picture below. A plant can possess both kinds of trend at different levels of height.

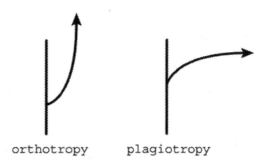

orthotropy   plagiotropy

The **order of an axis** is the number of axen to which the current one is attached to, down to the bearing axis.

Ramification is the process of branching. I.e. the divergence of the trunk into branches.

Phototropism describes the phenomenon of the bud growing towards light.

Branching can be attributed to three main categories: **monopodial, sympodial** and the rarer **dichasial**. The first type applies if a single continuous main shoot forms the stem of a plant, the second is given if different generations of apical shoots form the main stem. This is the cause when the apical bud of each of these sympodial branches has died, e.g. as result of fruit or flower germination. When two sympodial branches sprout at the same time branching is called dichasial.

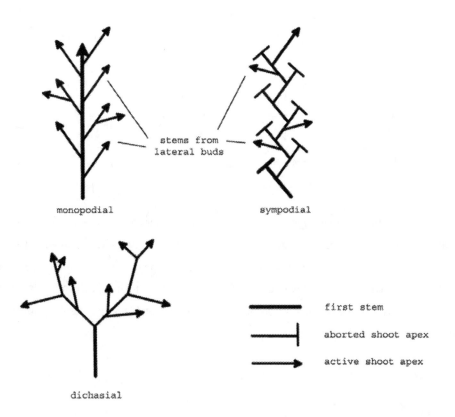

monopodial

stems from — lateral buds

sympodial

dichasial

first stem

aborted shoot apex

active shoot apex

3. Simulation

3.1 Introduction

As seen in the introductory chapter there exist a whole lot of different approaches to model plants, the choice for this attempt, however, fell on a simulation (in the spirit of De Reffye and Benes). While developing a complete simulation program is a daunting and time consuming task, modest goals and an extensively shortened wish list seemed to make this undertaking a promising journey into the realm of flora. To make a beginning the clue factors that are decisive in the definition of shape of a plant had to be sieved out and ordered according to their relative importance. Some of them, like a plant's phototropism, are quite obvious while others had to be confirmed and evaluated by experiment. In no particular order the following factors were deemed crucial in the development of a plant:

phototropism, phyllotaxis, branch growth rates, branching angles, branching type(sympodial, monopodial), branch mortality.

The resulting simulation is founded upon them, and moreover these factors are treated for the purpose of this thesis as a complete outline of plant growth. Many important factors have been omitted, because they either went beyond the scope of the goals of this thesis (e.g. wood density[24]), or would have consumed too much time to implement (e.g. bark[25]). However it is the strong conviction of the author that acceptable visual results with such a compilation are possible.

3.2 Light

Light has many effects on plant growth. On behalf of the amount of light intensity the plant is growing new shoots or determining growth speed as well as direction of these shoots. To incorporate this process into a plant model, estimates concerning the light intensity must be available.

Previous attempts to approximate that information divide themselves into three major categories. The use of extended radiosity functions[21], ray casting[2] and voxel spaces[22].

Radiosity is a costly method to compute the energy transfer between all polygons of an enclosed scene. Ray casting methods shoot virtual photons on the surfaces of the scene (or the other way around). If enough such photons have been emitted each surface should have a number of them assigned to it giving an estimate about the incoming light in that area. Voxel spaces are three dimensional regions partitioned into identical elementary cubes. These cubes can hold diverse types of information but are commonly used to describe occlusion, color or amount of light of a particular voxel.

In the original work of Green who introduced voxel spaces to plant growth simulation in 1989, he proposed to cast rays a certain amount of times into the sky and unto a 180 degree arc representing the sun's trajectory. The resulting number of unoccluded rays is used as a base for the amount of light that is available for that branch. So e.g. if for 100 rays that were cast 60 hit the 180 degree arc the resulting exposure would be 0.6. The model was later extended to account for the translucency of leaves.

In this work voxel spaces are used as well, but the casting of rays has been forsaken. Instead it is assumed that each voxel is maximally exposed to sunlight from the beginning. Then as the voxels become occupied by tree parts the value is decreased in relation to the size of the tree part. Additionally the values in the neighboring lower voxels are recursively decreased but to a lower extend as an attempt to simulate shadows. This is a very crude approximation, because modeling the actual physical processes would take an extraordinary effort. Each tree has thousands of leaves which reflect and absorb light in all directions. But only certain wavelengths of light are suited for the photosynthetic process, and only these are absorbed, so not only the intensity of light is altered but it's quality, too. On top of that, doing these calculations a single time wouldn't suffice because the sun travels the hemisphere over the course of a day in a nonlinear way. Even worse the trajectory changes slightly from day to day, where each day is subdued to varying climatic conditions. On the other hand plant species grow at different times of seasons in different intervals, which also has to be taken into account. All these arguments left no doubt that a simpler approach had to be implemented.

After making all these considerations the goal was set to grow a single tree in the open taking heliotropy and static external occluders into account. As mentioned earlier voxels are used here to hold information about sunlight exposure, and moreover are used exclusively for that purpose. The size of such a voxel cube is a little more than the average tree part at creation time. Each tree part can occupy only a single voxel, as determined by it's center of gravity. It is also possible for different tree parts to occupy the same voxel at the same time. Whenever a new tree part is added, the corresponding voxel decreases it's own exposure value and the values of voxels beneath it. This is done for all lower voxels that lie inside a conically shaped area up to a depth, where the exposure coefficient becomes negible. Algorithm 3.2, written in pseudo-code further illustrates the idea:

Algorithm 3.2: LightExposureUpdate()

```
    //Initialization of all four corners with the x and z coordinates of the occupied voxel
 1  corners1.xz=cubePosition.xz;        corners2.xz=cubePosition.xz;
 2  corners3.xz=cubePosition.xz;        corners4.xz=cubePosition.xz;
    //Find the height of the descent that is necessary
 3  while(lightExposure > lightExposureThreshold; i++)  {
 4          lightExposure /= reductionStep;
 5  }
    //Update all cubes below the occupied voxel down to descent height stored now in i
 6  currentLightExposure = lightExposure;  //this is a temporary variable
 7  for(k=1 to i){
 8          currentLightExposure /=  reductionStep;
 9          cube[cubePosition.x][cubePosition.y - k][cubePosition.z] += lightReduxTmp;
10  }
    //Repeat the above for all voxels neighboring the last plane of voxels
11  currentLightExposure = lightExposure;  //this is a temporary variable
12  for(k=1;k<i;k++){
13
14          corners1.x -=  1;        corners1.z -=  1;
15          corners2.x += 1;        corners2.z -=  1;
16          corners3.x += 1;        corners3.z += 1;
17          corners4.x -=  1;        corners4.z += 1;
18
19          diff=corners2.x - corners1.x;
20          for(j=0;j<diff;j++){
21                  for(l=1;l<i;l++){
22                          currentLightExposure /= reductionStep;
23                          cube[corners1.x + j][cubePosition.y -1][corners1.z]+=currentLightExposure;
24                  }
25          currentLightExposure = lightExposure;
26          }
27           diff=corners3.z - corners2.z;
28          for(j=0;j<diff;j++){
29                  for(l=1;l<i;l++){
30                          currentLightExposure /= reductionStep;
31                          cube[corners2.x][cubePosition.y - 1][corners2.z +j]+=currentLightExposure;
32                  }
33          currentLightExposure = lightExposure;
34          }
35           diff=corners3.x - corners4.x;
36          for(j=0;j<diff;j++){
37                  for(l=1;l<i;l++){
38                          currentLightExposure /= reductionStep;
39                          cube[corners3.x - j][cubePosition.y - 1l[corners3.z]+=currentLightExposure;
40                  }
41          currentLightExposure = lightExposure;
42          }
43           diff=corners4.z - corners1.z;
44          for(j=0;j<diff;j++){
45                  for(l=1;l<i;l++){
46                          currentLightExposure /= reductionStep;
47                          cube[corners4.x][cubePosition.y – 1][corners4.z - j]+=currentLightExposure;
48                  }
49          currentLightExposure = lightExposure;
50          }
51  }
```

3.3 Phototropism

Plants exhibit the tendency of growing towards light. This is necessary because photosynthesis cannot take place without certain amounts of irradiance available. And this process in turn is responsible to accumulate further mass. Up to a threshold this dependency on irradiance and growth is proportional[21]. The distribution of additional growth matter on the other hand is a dynamic process that still isn't very well understood. Concepts to model the process as an interaction of 'sources' (areas of carbon production) and 'sinks' (plant organs in need of matter) date back into the 60's. Although researchers are at disagreement concerning their role, it is certain that factors like proximity to carbon sources and type of plant organ are decisive in the allocation of matter[23].

For the purpose of this simulation voxel spaces are of particular use. As already explained above each voxel holds an exposure coefficient, or in other words a measure of incoming light. A branch can compare the values of nearby voxels to determine where the best light conditions are available. Furthermore, because growth is proportional to light this coefficient is used to determine the overall speed of growth of the branch, as well. Concerning the source sink interaction, the model proposed here reduces it to a user defined scaling factor controlling the relative allocation of matter between trunk and branches.

The direction of growth is calculated in the following way. First it is determined what cube face the tip of the branch is pointing to. Then the values of that voxel and the 8 voxels around it are looked up and vice versa for the next 9 voxels in pointing direction. To these values four different randomized variables are added, and the lowest value is chosen as the most favorable direction of growth. The four randomized variables represent the strategy of expansion of that particular plant. They describe the urge of growing directly upwards, holding the current direction in spite of any environmental conditions, growing horizontally or downwards. After the direction has been determined the next tree part that will be added to the top of that branch is going to be rotated towards it in an amount as defined in a separate variable representing the sensitivity of the phototropic reaction. The algorithm can be found below.

Algorithm 3.3: FindBestLightConditions(planeNormal, TreePart)

```
1  calculate scalars of planeNormal and all cube faces of TreePart.voxel;
```

```
2   chosenCube = choose normal of cube face with minimal scalar value;
3   vector.xyz = chosenCube.normal.xyz;
4   for (all 9 voxels in vector direction)
5          voxel.lightExposure += strategyValue + randomNumber;
6   chosenCube = choose voxel with maximum light exposure;
7   vector.xyz = direction towards chosenCube;
    //maxAngle is a parameter simulating intensity of phototropic reaction
8   rotate planeNormal towards vector in maxAngle amount;
*algorithm simplified
```

To give an impression of how the simulation procedures proposed above make themselves earned, an example has been prepared. Two trees that are situated close by start growing simultaneously. Several stages of growth were selected that show the trees without leaves, so that special attention can be paid to the growth directions of the branches. Note that no functions are present that control the growth direction of the branches. What can be witnessed on the next page is purely the result of the simulation itself. Although branching angles are used to initialize the direction of growth, the dominant impact can be attributed to the interaction of phototropic strategy and light conditions.

Development of two adjacent trees under
favorable environmental conditions

Iteration 10

Iteration 16

Iteration 25

Iteration 32

Iteration 39

Iteration 45

Iteration 45
Front with Leaves

Iteration 45
Back with Leaves

Special attention should be paid to the changes in branch development trends. While the older branches in the lower part of the tree are tending towards plagiotropy the younger branches exhibit orthotropical development. Furthermore the close proximity of the two trees noticeably reduces shoot growth on the sides that are facing each other. Most branches grow outwards which gives the impression of both trees having a single ovally shaped crown. The last image on the bottom right depicts the same tree as to it's left but from a different viewing angle. Note that all images presented in this work were created without shadows, which diminishes the aesthetic appeal somewhat. There are two reasons for this. The first one is that no such algorithm was available to the author at the time of writing. And the second is that due to the huge number of triangles used to draw the models the incorporation of such techniques wouldn't make real-time rendering possible.

As mentioned in the last chapter the four parameters responsible for the growth strategy of the plant, are important factors in defining the shape of a tree. The example above uses only a single set of parameters. Other examples with different strategies will be shown later but for the time being the other simulation functions are introduced.

3.4 Growth

The growth of a tree is realized in three different stages. One function is modeling longitudinal growth of internodes, another their radial growth and the last controls creation of new internodes. The latter will be explained in chapter 6, as it forms part of the main growth loop. A brief description of the other two is presented here, starting with the function that models radial growth.

In the accompanying program each internode is described by a set of 3D points called vertexes. In the final stages of the drawing process these points are the key information for the actual drawing pipeline. A special function is attached to each vertex that models it's spatial expansion over time. In respect to what kind of tree is to be modeled these functions can be modeled differently. The proper definition of these functions is a crucial step towards realistic tree models. They provide a very powerful and versatile tool, that has a huge amount of possibilities. But apart from such tree specific definitions the growth functions also have to account for two other very important factors in radial growth. Firstly they have to reflect the intensity of incoming light utilized by the photosynthetic process, this is done as already described in prior chapters, and secondly they have to model believable matter allocation.

The *pipe model*, which was introduced by Shinozaki et al. In 1964, reasoned that every leaf needs at least one pipe transporting water from the base of a tree[34]. This implies that the girth at a branching section must be wide enough as to contain all pipes that are going to leaves of child branches. It can therefore be concluded that diameters of two outgoing branches must match the diameter of the branch before them. This can be formulated in the following way,

$$F_0 = \sum_{i=1}^{n} \beta \cdot F_i$$

where n is the number of outgoing branches, F_0 the area of the parent branch, F_i the areas of the child branches and β a scaling control parameter. Such a parameter turned out to be necessary as a simple sum of areas didn't suffice in giving uniformly good results. In the accompanying program in relation to branch order, distinct values for β can be selected.

Other approaches are employed, most of them based on the observation of Murray[36] from 1924,

$$r_0^p = \sum_{i=1}^{n} r_i^p$$

where r is the radius of a branch and p a parameter that ranges typically from 2.49 for large trees and 3.0 for small ones[24]. For both methods presented, it becomes clear that the time complexity is quite low.

The modeling of longitudinal growth on the other hand is in respect to computation time a lot more expensive. Consider that on an elongation of a tree part (internode) situated at the bottom of the tree almost all other tree parts must be translated accordingly. If in each iteration all tree parts are to be elongated time complexity runs at $\Theta(n^3)$, where n is the number of tree parts. In order to optimize the process, each tree part holds information about all outgoing child branches. The reiteration algorithm has been implemented recursively and is given below.

Algorithm 3.4: updateLength(treePart, k)

```
   //update all tree parts higher then k
1  for(g=k+1; g<number of internodes in current branch) {
2          //call function recursively for each branch attached to treePart
3          for(i=0; i < number of children) updateLength(list[child[i]], -1);
4          translatePosition(list[g], elongation);
5  }
```

One last note on the function modeling radial tree growth. Although in most experiments a steadily increasing radial growth was modeled (a static growth rate parameter was employed), the implementation allows for a much wider variety of drawable forms. This is due to the fact that each vertex may be binded to a different growth function. In complex cases however the proper definition of such functions can become quite a difficult undertaking. In order to see how different the results can become consider the (weird) floating rock-like structures below, that were created using a modified growth function.

3.5 Branches

While much about the growth of branches has already been said, in this chapter several problems which were previously left unattended are discussed. Namely branch growth rate and ramification.

Botanical research suggests that the proportion of terminal branch growth to lateral branch growth has a huge impact on the overall shape of a tree. Sympodial trees have faster lateral growth speed than terminal ones, whereas the opposite is true for monopodial trees[35]. In the chapters concerning phototropism it has already been made clear how light has influence on speed of growth. This means that with the inclusion of a proportion parameter, the program owns now two parameters controlling the speed of branch growth. The results that can be attained by altering the proportion parameter (this refers to the proportion of terminal to lateral branch growth) include a large number of known tree shapes, some examples that were modeled are shown at the end of this chapter.

Whether and where a branch starts growing is evaluated in two separate functions. The first one decides where on the axis to place a bud. This is done by evaluating the light exposure values of nearby voxels and on account of the alignment preferences of the plant. These patterns are defined by determining how many buds are created simultaneously at one tree part and the angle between them. The same angle applies when creating buds for any tree part. It is added modulo 360 degrees to the value of the last tree part's angle and to an selectable offset. E.g. to model a distichous branching pattern the angle would be set to 180 degrees, the number of buds at each procreation step to 2 and offset to zero. Recall chapter 2 were a few number of patterns were presented. Note that phyllotactic patterns of leaves are modeled in the same way, but leaves are allowed to be placed at the tip of branches. Then after buds have been placed, in the following iterations of the main loop, the tree is able to sprout a new branch. This is likewise decided taking exposure values of nearby voxels into account. The procedure is analogous to the one given in 3.3.

On the following few pages the reader finds a compilation of tree models generated by the accompanying program. They are not faithful representations of particular species found in nature, but were created with the idea in mind to show what spectrum of shapes can be modeled. On the second page a group of tree models has been selected, that have rather strong lateral branch growth, on the third some sympodial trees are depicted, and finally on the fourth page models that exhibit dominant terminal branch growth.

Solitary trees with different
growing strategies. These tree
models and all other models in
this book are shown without
shadows.

On this page a selection of sympodial
trees has been prepared. All trees
were modeled with a different set
of developmentstrategies, but
under the same environemntal
conditions.

Monopodial trees, ranging from the
weird one below to more natural ones
on the right were chosen for this page.
Note the variety of tree crowns,
that are again the result of quite
different development strategies.

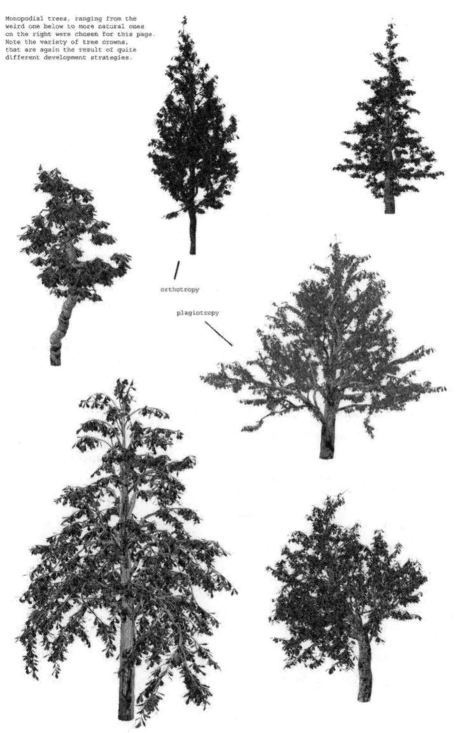

orthotropy

plagiotropy

gravitropical influence

4.Environment

4.1 Overview

While in the chapter dealing with phototropism the environment as a key factor influencing the growth of a plant already appeared. Here it's role will be expanded in such a way as to account for the quality of soil and light. Moreover the use of fuzzy controllers is employed in order to handle and simplify the large amount of parameters that can be measured in the underlying biological processes. For example the measuring of soil quality includes the measuring of particular nutrients, the acidic level and the amount of water. To model all these parameters separately would go beyond the aspirations of this work. This is regrettable because the respective deficiencies in soil quality are accompanied by sometimes completely different symptoms. E.g. on a shortage of zinc the younger leaves of a plant turn chloritic, whereas on a shortage of potassium leaves turn first dark green then purple and finally yellow. Such details are not modeled by this simulation, instead the mean of a variety of common symptoms is taken to represent a degree of health. This includes in particular, the visual representation of chlorosis and necrosis as well as terminal bud death and stunted growth.

To be able to model such environmental reactions, a suitable set of parameters has to be introduced. While a soil and a light parameter were obvious choices for the task at hand, additionally two parameters were selected to make genotype-environment interactions possible. Referred from here on as the soil and light dependency parameters. A weak dependency denotes that the plant is able to survive even under harsh conditions, but doesn't especially thrive under good ones. A strong dependency on the other hand allows for a fast growth but in good conditions only.

From these parameters a health degree is deduced utilizing fuzzy rule based systems as proposed by Mamdani[26]. Traditional rule based systems or the use of functions would be alternatives, but the inability to model uncertainty in the former and a tiresome nonintuitive adjustment process in the latter rendered them unattractive choices.

Once this was settled work began to formulate the details of the controlling system. A layered approach was conducted that in the first phase calculated sunlight and soil coefficients out of the aforementioned parameters. Then these in turn were used in the next layer to receive a general health variable leaving us so far with three output values of the inference process. These defuzzified output values are integrated in the main growth program through the employment of a set of crisp (nonlinear) scaling functions that compute the necessary growth model parameters (recall earlier chapters). These functions are not to be confused with the consequent of the inference process of the TSK fuzzy rule based system[27], where the rule consequents are also crisp functions as opposed to linguistic variables in the Mamdani approach. The model is illustrated below, and hardly needs further explanation safe for the mention that all the variables in the white boxes represent crisp values and not fuzzy ones.

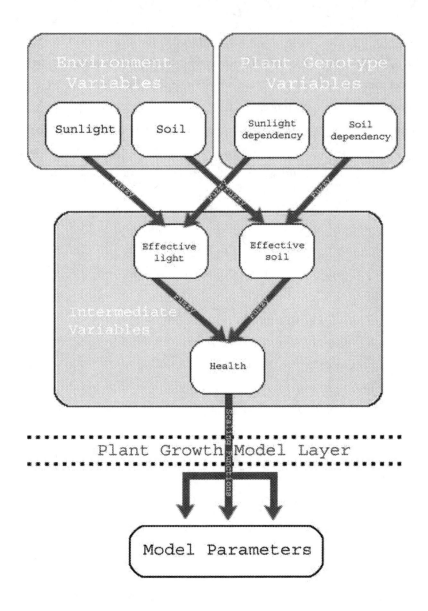

As can be observed above, the three variables Effective light, Effective soil and Health are logically grouped together in a box labeled mysteriously Intermediate Variables. They differ from the other variables in that they do not represent graspable natural phenomenona, but abstract ideas instead. They stand for the myriad scientifically measurable parameters that are actually involved in plant development and are dependent on the other two logical blocks. Although this seems as an overly simple model it is sufficient to introduce the desired visual gadgets that will give the impression of present environment conditions. In addition note that with the introduction of such a model it becomes possible to set up a selection process. The intermediary variables can be treated as fitness values, on which account a selection of genotype variables can be obtained. Later a simple version of an evolutionary algorithm will be proposed that will try to find the best possible configuration of genotype variables for a given set of environment variables.

The growth simulation parameters (at the bottom of the picture) used here are leaf color, leaf production frequency, growth speed and two parameters determining shoot production frequency. With decreasing values for Health the leaves would first turn to a yellow color and then to a brownish color to give the impression of chlorosis and necrosis, respectively. The Growth speed parameter accounts for stunted growth symptoms and finally shoot production frequency models occasional terminal bud death as well as overall shoot activity. All these parameters are computed with the help of scaling functions, that take intermediary variables as input values. As an example the reader finds the scaling function for leaf color presented below.

Scaling Function for the Leaf Color parameter:
Red = −((1.0 − (Health − 0.5)) * 1.9*(1.0 − (Health−0.5)));
Green = −((1.0 − (Health−0.5)) * 1.8*(1.0 − (Health−0.5)));

```
if(Red >= −0.7)       Color[0] = max( Normal(−0.7 − Red, var),−0.7 − Red );
else                  Color[0] = Normal(0.7 + Red, var);
if(Green > = −0.5)    Color[1] = max(Normal(−0.5 − Green, var), −0.5 − Green);
else                  Color[1] = Normal(0.5 + Green, var);
Color[2] = Normal(−0.8, var);
```
The color range is inversed and it's range is from 0 to -1. The function above named Normal refers to the Gaussian distribution function known from Probability Theory.

4.2 Fuzzy System

To set up a complete fuzzy rule based system, four different components are necessary. A module that fuzzifies the input variables also known as a data base. A so called inference engine that together with a fuzzy rule base module computes a fuzzy result set. Rule base and data base are often grouped together under the term knowledge base. And a module that defuzzifies the obtained set of the inference procedure into a crisp value. In this chapter one of the few fuzzy systems that were actually tested will be explained in detail. The program that accompanies this work was implemented with the idea in mind to make the creation of a user constructed fuzzy system as easy as possible. It can be heavily customized, and allows among other things the creation of norms, membership functions and rule base logic.

In order to fuzzify the input variables (soil, light, soil dependency, light dependency, effective light, effective soil) their ranges have to be known *a priori*. As in this case these values can not be measured in any way but instead are derived from a whole variety of different sources, their ranges were all arbitrarily set from [0,..,2]. With 0 being lowest and 2 highest reachable value. The linguistic states attached to these variables were defined as follows: very low, low, medium, high, very high. So for example, a high quality soil would pertain to a moist and rich in nutrients soil, although minor shortages on isolated minerals could apply. Here it has to be noted that more linguistic states are clearly preferable; in the opinion of the author from seven onwards. As the main interest lay in simple visual effects of plants in various health conditions, only five states were employed. This resulted sometimes in a coarse transition of inference results for antecedents that were in fact very close together. But this wasn't a major concern as a strictly faithful representation of the underlying biological processes never could have been an issue here. However, the program itself supports extensive customization, and therefore in the future can easily be refined according to the desired level of biological correctness. In chapter 6 the reader will find more information regarding this.

The linguistic states were represented by triangular shaped membership functions. Although experiments with s-shaped and Gaussian functions were also conducted the results differed only marginally. Users of the accompanying program are able to create their own membership functions but need to specify a name, body and range for it. All functions and data structures from math.h and tree.h libraries is available and random number generation is supported, too. For exemplary purposes the complete range of membership functions of the soil variable, as appeared in most of the experiments, is depicted below.

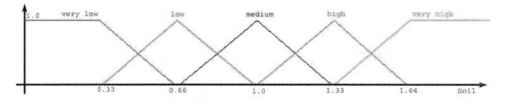

The next step is to define the inference engine and the rest of the knowledge base. In most fuzzy rule based systems the proper definition of the latter is the biggest obstacle. Hundreds of papers have been proposed, that deal with the incorporation of expert knowledge, but the prevailing occupation of scientists of the past two decades was the automatic derivation of the knowledge base. However, this was not the approach that was followed here, because a meticulous simulation of environmental processes wasn't one of the set goals. In addition recall that the values of the input variables are arbitrary numbers and the output variables constitute abstract values for visual effects of plant conditions which can be hardly measured. This makes the definition of evaluation functions that are necessary for any of the adaptive approaches a non-trivial task, at best.

As already mentioned the knowledge base can be logically separated into a rule base and a data base. Under a data base the description of linguistic variables, most importantly the membership functions, is understood. The fuzzy rules themselves compose the rule base. The image above already gave a glimpse into the oblivious manner in which the data base has been constructed here, and the rule base is likewise defined using the author's common sense only. Below the rule base that is part of the soil and soil dependency controller is presented.

S\SD	very low	low	medium	high	very high
very low	low	very low	very low	very low	very low
low	medium	medium	low	low	very low
medium	medium	high	medium	medium	medium
high	low	low	high	high	very high
very high	very low	very low	medium	very high	very high

The values inside the white fields refer to effective soil quality.

Traditionally these rules are formulated in if-else statements, which is a perfectly sensible way to do, when the fuzzy evaluation doesn't have to be done on the fly. Otherwise it is preferable to use static arrays, possibly with the aid of a smart indexing system. The reason for this is that modern microprocessors fetch and decode instructions into a pipeline before they are executed. While improving the speed of programs this also has a drawback. Whenever the code contains an if-else statement, the microprocessor doesn't know which branch to feed into the pipeline. Whenever the wrong branch is selected the microprocessor wastes valuable clock cycles to recover from the misprediction. The recovery time of AMD and Pentium M processors is approximately 12 clock cycles on a Pentium 4 it takes around 25 clock cycles[28].

With the knowledge base defined this still leaves the inference engine lacking. An inference engine is simply the selection of fuzzy operators responsible to making inference. In the accompanying program they are chosen prior to the start of the fuzzy evaluation process in a separate panel. Minimum and algebraic t-norm operators as well as the maximum t-conorm operator are standardly defined in the current version as of June '07. The user has the possibility to create his own norm or conorm by specifying it's name, type and body.

The last step of setting up a complete fuzzy rule based system is to choose a method that can deffuzify the fuzzy result set into a crisp value. From amongst the various approaches found in literature the center of area method was selected.

$$y_D = \frac{\int\limits_{-\infty}^{+\infty} y \cdot \mu(y) \cdot dy}{\int\limits_{-\infty}^{+\infty} \mu(y) \cdot dy}$$

$\mu(y)$ is the fuzzy set of the inference process (consequent). The technical realization included the utilization of Riemann integrals with a granularity of 50 that operated on lookup tables of the consequents. The method yielded as expected overall satisfactory results, but one oddity was observed during the experiments that deserves special mention. Because the method does not take the size of the integral of $\mu(y)$ into account, the value for y_D stays the same for different antecedent pairs having the same consequent when only one rule fires. Consider the following situation.

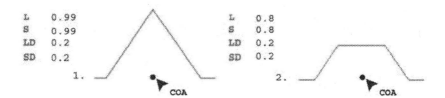

For both inference result sets the value obtained is exactly the same. However it would be desirable that the value in the second case would differ somewhat from the first. This can be remedied when the number of membership functions is increased and the rules accordingly refined. The inherent flaw will of course remain but such situations will arise less frequently. Alternatively the defuzzification method could be improved by taking e.g. the size of the integral into account or choosing a different method altogether, but no such improvements were deemed necessary for the purpose of this work.

On the next page the reader finds a selection of images of two kinds of trees that were created using the accompanying program. The model parameters were in both cases exactly the same, so the results were achieved by changing the environmental and genotype variables only.

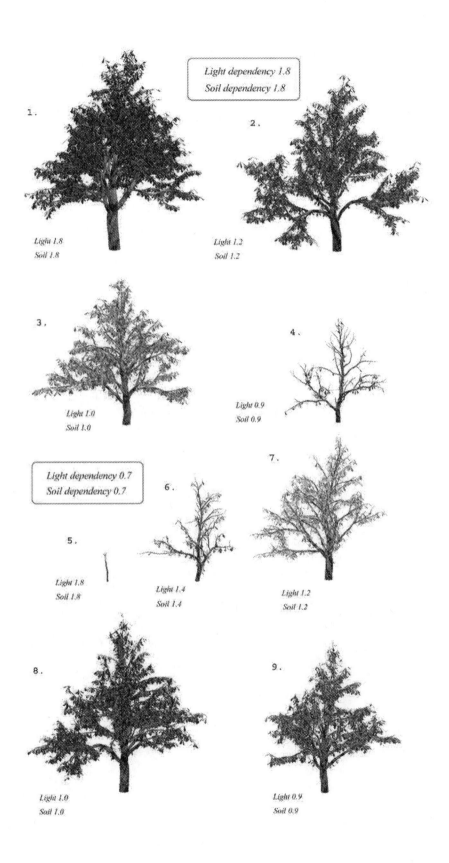

1.

Light dependency 1.8
Soil dependency 1.8

Light 1.8
Soil 1.8

2.

Light 1.2
Soil 1.2

3.

Light 1.0
Soil 1.0

4.

Light 0.9
Soil 0.9

Light dependency 0.7
Soil dependency 0.7

6.

7.

5.

Light 1.8
Soil 1.8

Light 1.4
Soil 1.4

Light 1.2
Soil 1.2

8.

Light 1.0
Soil 1.0

9.

Light 0.9
Soil 0.9

With high soil and light dependency values, the upper tree is much more suited for favorable environmental conditions. Therefore in picture one a lush and thriving tree has been modeled, while with decreasing environmental conditions more and more deficiencies become apparent. First inhibited growth, fewer branches and a slightly greener leaf color occur, but in picture two and three are rapidly moving on to severe chlorosis and necrosis displaying extreme stunted growth symptoms. In picture five the second tree on the other hand, does not grow under the same, extremely good conditions of picture one altogether! Only as the intensity of light becomes smaller the tree recuperates little by little reaching it's full potential under moderate conditions. Also notice that although in picture nine the second tree does grow less vigorously than the picture before, it is still much more suited to the harsh conditions than the first tree.

5. Evolutionary Algorithm

5.1 Vision

The proposed model of genotype-environment interaction can be further expanded with the addition of a selection process. The intermediary variables serve as a measure of fitness of the genotype variables for a concrete set of environment variables. Concepts like selection and reproduction are known and applied in computer programming for quite some time now. Actually from the very beginning of computer science did biological inspiration serve to construct algorithms. Pioneers like Turing or von Neumann, who envisaged among other things the use of cellular automata capable of reproduction, already showed interest in an adaption of the biological notions observed in nature. The first concrete ideas were formed in the early sixties when evolutionary programming[30], genetic algorithms[31] and evolutionary strategies[32] were introduced. These methods belong to the family of heuristic optimization and search techniques that nowadays are grouped together under the term evolutionary algorithms. The basic procedure of such an evolutionary algorithm is any case the same. First a starting population of possible solutions must be defined. Out of this population individuals are chosen to procreate with other members of the population. The use of genetic search operators like mutation and recombination is employed to define the traits of the resulting child population. An evaluation process determines a fitness value for each population member, and a selection process 'culls' the population of the next generation. To receive a more in-depth look into the respective differences between the three main methods of evolutionary algorithms the reader is inclined to take a look into the corresponding literature[29].

In this chapter the utilization of evolutionary algorithms in plant modeling is discussed. Alongside with it a particular implementation of such a procedure is presented. The reader should note that this chapter serves merely as an outlook and general outline for the problem.

As portrait in the last chapter four distinct parameters are used in the accompanying program to simulate a genotype-environment interaction. Assuming the parameters for light and soil are static, that is to say describe a stable unchanging environment, it becomes quite easy to incorporate evolutionary principles. The optimization domain becomes the fitness landscape as defined by the real-value genotype parameters. The measure of fitness of each parameter set is simply the value of the Health variable. The nature of the problem (parameter optimization) invites to choose the evolutionary strategy approach. To avoid confusion it's worth noting here that this is different to most genetic fuzzy rule based systems, were typically the adaption capabilities of genetic algorithms are utilized to construct a rule base or data base[33]. Here evolutionary algorithms are used strictly for parameter optimization purposes. Although for a problem of such complexity as the one presented here traditional optimization methods from numerical mathematics would suffice too; the problem's specific nature allows to make profit of a previously unexploited trait of evolutionary algorithms. As stated earlier, populations of various generations are used as sets of domain solutions that converge towards a global optimum. This mechanism is exploited so far only in respect to it's functional nature; for the problem at hand, in addition the populations themselves become subject of interest. i.e. information like composition, size and mutual proximity of these populations can be used as a basis to create entire plant model populations under given environmental conditions. To further illustrate the idea, an example is presented below.

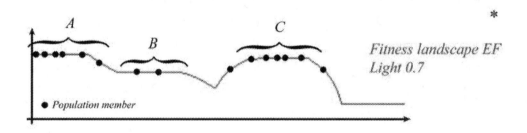

*rule base logic was slightly modified

Let us assume that the black dots denote fitness values for specific members of the population of a particular generation. Due to the relative proximity of population members, a logical grouping into three partitions A, B and C could be attained. These in turn could serve as a base for our plant model population. Earlier it was shown how health characteristics like leaf color or growth speed can be computed based on the intermediary variables. In the same manner other plant characteristics can be connected to the process. Lateral branch growth speed and phototropic strategies could be included into the set of genotype variables, while being in part dependent on the already introduced genotype variables. That way the respective members of the three populations would exhibit common visually perceptive traits among themselves, but quite different ones compared to other populations. E.g. population A could resemble a more coniferous tree while population C a deciduous tree type.

It becomes immediately apparent that certain additional constraints and modifications have to be introduced to the typical evolutionary strategy. But first some insight into the method itself, there are several versions of it described in literature. They become more and more complex, beginning with the two-membered approach $(1 + 1)$ on to the multi-membered approach $(\mu + 1)$ that produce only a single offspring replacing it's parent in each iteration, were μ is the size of the population. The latter evolved into the $(\mu + \lambda)$ strategy, with λ referring to the number of offspring that is temporarily added to the population. A selection process reduces it to μ individuals immediately afterwards. And finally the (μ, λ) strategy with $(\mu < \lambda)$, were a new population of μ individuals is selected in each generation of the child population only. The latter is useful when the optimization problem is dynamic in nature, and high flexibility is a fundamental condition.

For the specific problem of creating plant model populations, only the latter three are worth considering. The (μ, λ) strategy operates a little too fast, because it is desirable to actually see the plants grow and eventually die. This leaves the $(\mu + 1)$ and $(\mu + \lambda)$ strategies which can be treated almost indifferently in respect to the specific implementation. As partly already stated the original approach performed selection in a deterministic way, by either replacing the corresponding parent with their offspring or by eliminating the λ least fit members. In the case at hand it becomes necessary to attach a mortality parameter to each of the members that can account for a gradual control of removal from the population. Another necessary modification concerns the procedure in which evolutionary strategies choose members to act as parents. Standardly both partners are selected with equal probability, which does not reflect botanical reality, were cross-breedings between different species are less probable to occur. This calls for a probabilistic selection of partners taking both their virility as well as relative proximity of genotype variables into account. However, this has consequences on the optimization capabilities of the method. By making partner selection based on genotype proximity, the recombination operator looses much of it's search functionality leaving only the mutation parameter as a tool to prevent premature convergence. Although evolutionary strategies began without a recombination parameter altogether, the community successfully adopted the idea lately from genetic algorithms[29]. To counteract this shortcoming free partner selection can be permitted from the beginning, but gradually made unwanted as the populations begin to converge towards particular optima. A full analysis of such a method is beyond the scope of this work but a concrete realization can be found in the next chapter.

5.2 Realization

Population members are represented by a vector (x, σ, α), where x is a vector of the genotype values, σ the standard deviation and α the age measured in generations of the member. Resulting offspring is calculated the following way, first a parent is selected with a probability of

$$p_a = \frac{x^{(a)}}{\sum\limits_{i=1}^{n} x_i}$$

where n denotes the size of the population. After this initial step, a partner is chosen in relation to it's proximity to the already selected parent. The formula of selection probability stays the same but the values for x_i are modified accordingly,

$$x_{new} = x \cdot \left(1 - \frac{\left|\sum\limits_{i=1}^{m} x_a^{(i)} - \sum\limits_{i=1}^{m} x^{(i)}\right|}{2 \cdot norm}\right)$$

where x_a is a already selected parent and x a member of the population. The variable *norm* refers to a normalization coefficient so that the values lie in the range of 0 to 1 and m is the size of the vector x. The number 1 stands for a vector with m ones as components. Once two parents have been selected the recombination operator is applied to create an offspring. Two types of crossover are employed in evolutionary strategies, known as discrete and intermediate recombination. In discrete recombination each vector component can come from either parent,

$$(x_{off}, \sigma_{off}) = ((x_q^{(1)}, ..., x_q^{(m)}), (\sigma_q^{(1)}, ..., \sigma_q^{(m)})) \qquad , q \in \{a, b\}$$

whereas in intermediate recombination the mean of two vector components is taken,

$$(x_{off}, \sigma_{off}) = ((\frac{(x_a^{(1)} + x_b^{(1)})}{2}, ..., \frac{(x_a^{(m)} + x_b^{(m)})}{2}), (\frac{(\sigma_a^{(1)} + \sigma_b^{(1)})}{2}, ..., \frac{(\sigma_a^{(m)} + \sigma_b^{(m)})}{2}))$$

which was preferable because only two component vectors are used in this case. Then a mutation operator is applied to the offspring ($\Delta\sigma$ is a parameter of the program).

$$x_{off}' = N(x_{off}, \sigma_{off})$$

$$\sigma_{off}' = \sigma_{off} \cdot e^{N(0, \Delta\sigma)}$$

Note that the deviation parameter also undergoes mutation resulting in self-adaption of control parameters. Other forms of mutation for the deviation are known, like e.g. the 1/5 success rule, where in relation to the number of successful mutations the parameter is either increased or decreased.

The last step of setting up an evolutionary strategy procedure, is to define the selection process. As already explained above in the (μ + 1) and (μ + λ) method the populations are reduced deterministically to μ members after each procreation step, according to their fitness values. This required further refinement to suite the needs of plant modeling. A special mortality function was introduced to account for a gradual removal from the population. Whenever offspring was produced the function evaluated the removal of the least fit member of the population. On a positive result (i.e. member dies) the offspring would occupy it's place whereas on a negative the population size was increased and offspring was appended to the end of the list.

$$f(x) = \begin{cases} 1 & x < \alpha \cdot \Theta \\ 0 & x \geq \alpha \cdot \Theta \end{cases}$$

Parameter Θ is the threshold of removal. On a higher value the chance that the population size increases gets larger. The parameter doesn't need to be static, as a matter of fact in the underlying problem, it would be preferable that it would vary from function call to function call. That way further compensation for the restriction imposed on the recombination parameter earlier can be achieved. At the beginning of this chapter it was stated that environment conditions are considered stable, but when the threshold parameter behaves dynamically a certain degree of variation applies. The algorithm in pseudo-code is presented below.

Algorithm 5.2: MakeEvolution(pop)

```
1  for(i=0; i<maxGenerations){
          //calculate the fitness values for all population members
2         for(j=0; j<popSize){
3              tmp1=calcEffectiveLight(pop[j].lightDep, light);
4              tmp2=calcEffectiveSoil(pop[j].soilDep, soil);
5              popFitness[j]=calcHealth(tmp1,tmp2);
6              popAge[j]++;
7         }
8         for(each parameter p){
               //choose parent according to fitness
9              a=chooseParent1(popFitness);
               //choose parent according to fitness and proximity to a
10             b=chooseParent2(popFitness,a);
               //create offspring using mean of parameters in pop array
```

```
11              child.p=Normal((pop[a].p+pop[b].p)/2, (dev[a]+dev[b])/2);
                //find most unfit member of population
12              min=Minimum(popEval);
                //on positive result replace the member
13              if(checkRemoval(min, popAge[min])){
14                      for(each parametr p) pop[min].p=child.p;
                        //mutate deviation
15                      dev[min]=dev[min]*exp(Normal(0.0f, deltaDev));
16                      popAge[min]=0;
17              }
                //on negative result append offspring to the end of the list
18              else{
19                      for(each parametr p) pop[popSize].p=child.p;
                        //mutate deviation
20                      dev[popSize]=dev[popSize]*exp(Normal(0.0f, deltaDev));
21                      popSize++;
22              }
23 }
```

6. Description of the program

6.1 Introduction

In this chapter a discussion of the implementation itself is presented. Naturally, not everything can be covered so the author opted for a choice selection of the most interesting parts. Among them is the proper introduction of the main growth loop as the heart of the application. As the reader may recall, in the chapter amending plant growth one aspect has been deliberately omitted, because the author believed it more sensible to place it along with the implementation. In this case the idea is so strongly connected to the technical realization that an earlier discussion of it would have been quite cumbersome.

6.2 DLL methods

To allow for a convenient modeling of the aforementioned plant processes the use of run-time DLL linking was employed. Each process was assigned a special function in the DLL, that was solely responsible for the proper simulation of that particular problem. In that way the whole simulation process was separated from the graphical representation and the main plant growth loop of the program. In a script like manner the function body along with it's mass on parameters could be adjusted and tweaked without the need for recompilation. Thus the program receives a highly didactic component, because future researchers experimenting with it are able to easily introduce and test their own approaches, without bothering to implement the rendering part or the graphical user interface.

The technical solution to achieve that, included the use of the license free console compiler from Digital Mars. It compiles and links a CPP file that holds all the user defined simulation functions into a DLL when a button on the GUI is pressed. These are later binded to function pointers inside the program once the main loop is started. The body of every simulation function can be loaded into an edit box and altered in any way that the C++ language with the math.h library and a program specific header (tree.h) allows. Any changes become immediately applicable to the program.

6.3 Main Growth Loop

The following brief summaries of program classes are indispensable for the proper understanding of this chapter.

- ## class Tree

The most fundamental of all classes. It owns a static TreeBranch array named list. All TreeBranch objects later created must be added to that list. (Note that all arrays used here are static arrays. This is a decision motivated by the notion to obtain a fast and efficient application. While dynamic arrays would be in this particular case desirable, the speed bump caused by them by far outweighs their advantages.) An unsigned integer variable n holds information about the current amount of TreeBranch objects associated with the tree object. The most important of it's method is called the growTree function were the main plant growth loop is located. Lookup tables for trigonometric functions and a three dimensional array representing the voxel space also reside within it.

- ## class TreeBranch

The TreeBranch class mimics in many ways the already mentioned Tree class. But it holds TreePart objects instead of TreeBranch objects and has an additional TreePart static array named leaves, used exlusively for TreePart objects that represent graphically and functionally leaves. Apart from that the class holds information about the order of the TreeBranch in the treeLevel variable, the growth speed of the branch and a pointer to the parent TreeBranch object to which it is attached to.

- ## class TreePart

As we already know all TreePart objects find their way into the static array lists of the TreeBranch class. They form the elementary objects of our Tree hierarchy. Here are all detailed geometric and functional information of plant parts located. We find the number of possible buds, the geometric representation as a Part3D object, the part type and age, direction and position of the part and other useful information. In order to perform efficient updates of the plant geometry, the authors also included information about TreeBranch objects that go out of this TreePart. This may seem confusing, because it was stated earlier that TreeBranch objects create other TreeBranch objects, but these have to be positioned in three dimensional space and must have contact with a TreePart of that TreeBranch. This is the very TreePart that takes note of these outgoing branches for later computations.

• class Part3D

While the other classes in Tree.h represent a compilation of various kinds of information, the Part3D class is strictly a collection of geometrical data. We have static arrays containing the spatial positions of points in 3D and corresponding normal vectors of these points. The growthFunction static array of function pointers will later contain addresses of DLL functions. Each TreePart object must contain a Part3D object describing its geometrical representation.

For the sake of performance object hierarchy of the plant simulation is kept simple. A globally defined Tree object is created at the beginning of the winMain function. After the user started the Tree::growTree method, successively a TreeBranch object (the trunk) and a TreePart object (the root) are created prior to entering the main loop. The TreeBranch object is added into a static list, which is a public member of Tree. The same applies for the TreePart object, which in turn is added to the TreeBranch list. Each TreePart object constructor requires exactly one Part3D object. These objects are read from a static array which comes as a parameter of the Tree::growTree method. The program includes two ready made Part3D objects. A circle with variable precision and a diamond shaped leaf form. After all this initialization the Tree::growTree method enters the main plant growth loop, creating and placing further TreePart and TreeBranch objects into the appropriate lists, until the user decides to stop it.

Below a simplified version of the algorithm containing the main loop is presented.

Algorithm 6.3: PlantGrowth()

```
1 Seed random number library;
2 Initialize the voxel space with 0;
3 Initialize the lowest plane of voxels with an arbitrary high number;
4 create TreeBranch; //the trunk
5 create Part3D form; //a circle in our program
6 create TreePart; //the root
7 Load the DLL;
  //all simulation functions apart from the growth functions are binded here
8 bind DLL functions to function pointers;
```

```
       // the beginning of the main loop
 9   for(i..age){
10       for(j..number of tree branches){
             //returns destination direction, angle and speed of growth
11           call lightFunction;
12           if(growthSpeed > growthSpeedThreshold){
                 //described in detail in the chapter Geometric Representation
13               smooth the angle if necessary;
14               do stochastic variation of the direction;
15               create TreePart;
16               add TreePart to current TreeBranch;
                 //bind DLL growth functions to the function pointers
17               bindGrowthFunctions;
18           }
             //no further tree parts are added
19           else declare tree branch as dead
             //decide whether and where to grow buds
20           call markFunction; //from DLL
21           if(i==age) create Leaf; //code almost like above, not shown
22           for(k..number of tree parts in current tree branch){
                 // makes the spatial changes of the 3D points
23               call growthFunction for each Vertex; //from DLL
                 //increases the length of part and updates the rest of tree
24               call updateFunction; //from DLL
                 // kind of branch
25               if(current treePart.type==SHORT) remove it from the model
26               if(i==age) create Leaf;
                 // decides if a bud grows into a new shoot
27               if(call branchFunction==success){
28                   for(e..number of buds){
29                       if(bud shall grow now){
30                           create TreeBranch;
31                           add to Tree list;
32                           compute direction;
33                           create TreePart;
34                           add TreePart to current TreeBranch;
35                           bindGrowthFunctions;
36                       }
37                   }
38               }
39               current treePart.age++;
40           }
41       }
         // see below for a short description of the method that is resumed here
```

```
42    resume write2Arrays;
      //when the above method finishes it will resume this method
43    suspend self;
44 }
```

6.4 Program Workflow

Below a short summary of three methods is presented. In the following text several references will appear to them.

- **method write2Arrays**
  This method reads positions and directions from our tree object and saves them into static arrays. Additionally it also computes the bi-normal and tangent components as well as texture coordinates of each vertex. It iterates branch for branch tree part for tree part until finished.

- **method createPlant**
  This method creates two Part3D models (a circle and a diamond leaf form), then it calls the growTree method of that object and the feedback loop is underway.

- **method growPlant**
  First the method calls the write2Arrays and createPlant methods in two separate threads, after which it is halted by a closed semaphore. When the write2Arrays method opens the semaphore, the CreateGLWindow method is called and keyboard listening loop is entered. Upon exiting the thread handles are released and the OpenGL window is destroyed.

The major problem in embedding the plant growth simulation into the whole program, is to synchronize the communication between the OpenGL rendering pipeline on the one hand and the main loop in Tree::growTree on the other. A serial approach would also suffice but the plant wouldn't be drawn step by step (i.e. you don't see the plant grow).

The write2Arrays function reads data from the Tree object created priorly by the createPlant method. Both these functions are called in different threads in the growPlant method, which then in turn runs into a closed semaphore. Note that the write2Arrays function is created suspended, as the plant model must accomplish one full iteration before data can be read. The createPlant method creates a Tree object and invokes it's growTree method passing all necessary thread handles and semaphores along the way. After growTree's first iteration the write2Arrays function is resumed and growTree suspends itself. In the very first call of the write2Arrays method a semaphore is released so that the growPlant method may continue to create an OpenGL window and start keyboard listening. The feedback loop between write2Arrays and growTree is continued by the consecutive pressing of the 'Q' key. Now that the arrays holding vertex, normal and texture positions already contain data from the feedback process the OpenGL Renderer is started in a new window. It is synchronized with the write2Arrays function, because otherwise it would be rendering incomplete vertex data. The picture below illustrates the program workflow concentrating on the plant growth simulation part only. The content boxes hold specific method names but stand for a logical compilation of many methods actually involved, other than that the author hopes that the picture is self-explanatory.

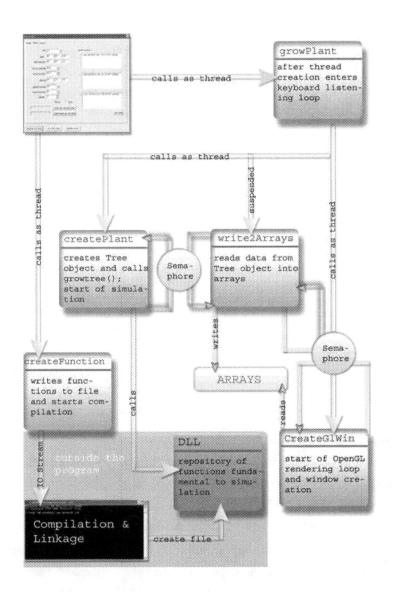

growPlant

after thread creation enters keyboard listening loop

calls as thread

calls as thread

suspended

calls as thread

createPlant

creates Tree object and calls growtree(); start of simulation

Sema-phore

write2Arrays

reads data from Tree object into arrays

writes

calls

Sema-phore

createFunction

writes functions to file and starts compilation

to Stream

outside the program

ARRAYS

reads

DLL

repository of functions fundamental to simulation

CreateGlWin

start of OpenGL rendering loop and window creation

Compilation & Linkage

create file

6.5 Basic Plant Growth Realization

As explained in the last chapter the plant growth simulation uses elementary tree parts as means to hold geometric information as well as functional data. The whole process of plant growth takes place in a loop that iterates branch for branch through the whole tree, creating new tree parts in each iteration. The position and other data is computed on account of the top last tree part of the current branch. Five distinct simulation function which are located externally in a DLL are called and evaluated by our main loop in each iteration. They try to represent functional processes of a plant considered most important in its growth. First a function simulating phototopism is evaluated, it returns a floating point number representing the strength of irradiance for this tree part as well as a vector representing the general direction in which the plant wants to grow. In respect to that floating point number either another tree part is added to the top or not. Then the main loop reiterates through our whole plant calling first a function simulating the creation of plant buds, then functions simulating radial or longitudinal growth of tree parts and finally a function that decides whether and where a bud turns into a new shoot.

6.6 Geometric Representation

As stated above, the program uses elementary tree part class objects as place holders for variables concerning the geometric representation and process data necessary for the simulation functions. Before the beginning of the growth model a set of 3D points has to be defined by the user. Typically, to model a tree, this could be a number of points along a normalized circle. All these points get a growth function assigned, i.e. a function that computes the x, y and z coordinates in respect to a time value. The growth function can take also other parameters into account, like branch order or light exposure. Whenever in the main loop a new tree part is created, the plane normal and the position of the last tree part is used as a base for the phototropic calculations which return a target normalized direction. The last tree part's normal is then rotated towards that direction in such an amount as a special parameter describing the plant phototropic reaction allows to do. To be able to realize the rotation, first the scalar product between normal and direction and then the (normalized) cross product is calculated, leaving us with a valid angle and axis. Instead of such an approach the use of quaternions could be considered, because they allow to do the same and are slightly faster. The overall gain in performance, however, wouldn't be that much better because the addition of tree parts isn't a major bottleneck. For comparison, the function computing phototropism runs at over 60,000 clock cycles while the whole tree part addition requires merely 20,000 clock cycles of processing time.

Assumed the new tree part's normal is already calculated, that still leaves the new tree part's position wanting. Therefore, along with the computation of the direction, the phototropic function delivers a parameter responsible for the speed of plant growth. And this in turn is used as a measure to translate the new tree part in plane normal direction. At this point it can turn out that the light exposure is so small that the addition of a new tree part becomes undesirable. In that case the tree part is not translated at all, and the tree part is not added to the static list of the current tree branch. Once that occurs the branch will never be able to add a tree part again, becoming dormant and breaking off of the tree after a set number of iterations.

Occasionally when the rotation angles turn out to be so big that the tree branches start looking unnatural, a special mechanism kicks in to counteract this. It keeps dividing the rotation angle until it is below a certain maximally tolerable amount, and then uses the number of times that were necessary to do that to create an equal amount of shorter tree parts. The result is a smoother transition of tree parts so that they look more faithful to botany.

6.7 Drawing Process

The geometric data held in a tree part class object consists of the following. The plane normal, the world coordinate position of the center of surface, the normals of all the 3D points, the world coordinate positions of all the 3D points, function pointers to growth functions of all 3D points and the position in voxel space of the voxel that this tree part is attached to. To actually draw a picture out of this information on the monitor the program makes use of the OpenGL Renderer. There are several possible methods available that actually draw the 3D point information, and all require a specific representation of the necessary data. Here, in respect to the tree parts representing branch parts, the glDrawElements method in GL_TRIANGLE_STRIP mode was utilized. It needs the positions and normals from all tree parts stored in either one or in separate arrays. The drawing algorithm reads triangles in the following fashion. First k, k+1, k+2 is drawn then k+2, k+1, k+3, after which it switches back to draw k+2, k+3, k+4 then k+3, k+2, k+5 and so on. In this way, after initializing the first triangle with three points, the addition of a single point is sufficient to draw a further triangle.

As already described, each tree part holds the geometric information of a set of points, which in most experiments meant usually 16 points aligned uniformly along a circle. Between all such adjacent circles of a branch the triangles are drawn in the manner described above. This has to be done for all branches of a tree separately. Below the order of the points stored into the arrays is depicted.

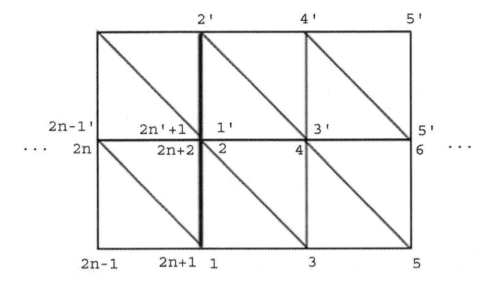

Note that when drawing 2n+1, 2n, 2n+2, the cylinder is closed, but the algorithm goes on to draw 2n+1, 2n+2, 1' and then 1', 2n+2, 2' which are in fact redundant lines (thick line). Then however the process is continued in proper manner and 1', 2' ,3' is drawn.

All vertexes apart from the first circle of a branch and the last circle are stored twice and in the case of the first vertexes of each circle four times (e.g. 2,2 n+2, 1', n'+1). The total amount of memory in bytes used for each branch is therefore: (2n+2)*(number_of_circles-1)*4.

6.8 Shading Effects

In addition to the positions of vertexes and it's normals, the program also computes the tangent, bi-normal and texture coordinates. The bi-normals can be computed by subtracting the points of the higher tree part and the points located below them (e.g. 1'-1) . Then the cross product gives us the tangent values. If the difference of growth between two tree parts is rather small, the bi-normals don't have to be computed, because their value can be approximated with the plane normal.

To enhance the visual aspect of the project a pixel and a vertex shader in Cg language were written. They contribute in two main ways to the overall visual aspect of the project. Firstly they introduce per-pixel lighting to the OpenGL Renderer. Standardly the Renderer only interpolates data from the normals of each Vertex. Although this produces viable lighting results it is imprecise and fails to convince in certain circumstances. In per-pixel lighting the normals are read from the normal map and lighting is computed for each pixel of that map separately. Due to the nature and huge amount of additional computations involved, it was necessary to do them on the graphical processing unit. Secondly they allow to bump map textures. An approach known as TBN-bump-mapping was used, which offered a fast implementation. TBN stands for Tangent, Bi-normal and Normal and refers to the orthonormal basis of the texture map. A tangent is the vector going horizontally along the texture, a bi-normal is the vector going vertically along the texture and a normal is the vector pointing out of the texture. In TBN-bump-mapping these three vectors are calculated beforehand and saved into arrays. The normals come as part of the simulation data, the bi-normals can be approximated with the plane normals of the tree part, and the tangents are simply the cross products of the other two vectors. They are passed to the shader as 3D texture coordinates, and translated into the necessary dimension. Light and eye position are passed as uniform parameters, that is to say from outside the shading unit, and also translated to the necessary dimension. Drawback of this approach is that, if the model is rotated or translated the TBN calculations have to be redone. Many different algorithms and implementations for bump-mapping exist, for a complete review of them the reader is inclined to take a look into the corresponding literature. Below a close up view of some bump mapped bark texture is presented.

without bump-mapping   with bump-mapping

close up of tree

6.9 File I/O

The program also offers the possibility to export and import the constructed plant model into the data format OBJ (object). Currently only the vertex positions and the order of of the drawing process are taken into account. For a full blown OBJ script the inclusion of normal directions, texture coordinations and leaf data would be desirable. But for the latter it is in most cases not needed as the modeling of real-time plants is still somewhat limited and faster approaches of visual representations have preference. The other two will be included in the future, especially as the program will generate a bark texture along with the growth simulation.

The applicable method to do all this is write2OBJ, which apart from writing relevant visual data also writes information concerning the number and length of branches into the comment part of the script. This additional information is so far only usable by the program's own OBJ parser implemented in the readFromOBJ and ensures a faster visual reproduction.

Other formats, especially 3DS will be supported in the future, allowing for more detailed scripts with the inclusion of texture data, bi-normal and tangent directions, etc.

7. Conclusion

The enormous task has been tackled. Although countless unamended problems connected with plant modeling still remain, the major goals have been reached. Along with it, new enthusiasm for previously unnoticed fields of research has been discovered. Especially the first steps in artificial life were unusually enjoyable to the author and fanned the flames to delve deeper into the realm of scientific wonders. The work itself stayed entertaining until the very end, with moments of epic lunacy and utter madness abounding along the path. And now, that this time-spanning undertaking has journeyed to it's end, the last chapter's sentences do not come out easy. So the author decided to spare the dear reader his unmanly sensitivity, stop the erratic babbling, and give what everyone really wants to see: *a jazzy tree*.

List of objects and methods

List of objects

- branchTypes
- Function
- FunctionList
- FunctionTriangle
- Fuzzy
- FuzzyLogic
- FuzzyPlantSystem
- FuzzyRulesSystem
- interFunctionList
- interFunctionListWrapp
- marker
- Part3D
- PlantgrowthTypes
- Rules
- SoilTypes
- SunlightTypes
- Tree
- TreeBranch
- TreePart

List of methods in "Tree.cpp"

- bindFunction(char * functionName, HINSTANCE hinstLib, TreePart t, int c, int a)
- cross(float * a, float * b, float * c)
- defuzzification(Function * f, int n)
- exFunc(float a)
- maxFuzzy(Function * a, Function * b, Function * c, Function * out, float start, float stop)
- maxFuzzy(FunctionList * fl, int k, Function * out, float start, float stop)
- minFuzzy(float a, Function * f)
- myAbs(float x)
- myAbs3D(float * x)
- myAbs3D(int * x)
- normalize(float * x, float * y, float * z)
- printPos(Part3D * p3d)
- printtp(TreePart * t)
- scalarProduct(float * vec, float * dvec)
- testFuzzyRules(int a, int b)
- testMark(TreeBranch * t, float * prob, int c)
- transPart(Part3D * part, float * pos)

List of methods in "ctlone.cpp"

- addInterFunction(char * functionName, char * functionBody, char * beginEndBody)
- addLogicFunction(char * functionName, char * functionBody, int mode)
- addNewFunction(char * functionName, char * functionBody)
- addNewLogicFunction(char * functionName, char * functionBody)
- compileDLL(char * functionName, char * functionBody, int mode)
- createDll()
- createFunction2(void * cp)
- CreateFunctionDialogProc(HWND hwnd, UINT message, WPARAM wParam, LPARAM lParam)
- createInterFunction(char * functionName, char * functionBody, char * beginEndBody, int mode)
- CreateNormDialogProc(HWND hwnd, UINT message, WPARAM wParam, LPARAM lParam)
- createNormFunction(char * functionName, char * functionBody, int mode)
- createWrapper(interFunctionListWrapper * iFLW)
- decreasingTrapezTest(float x, float * p)
- Dialog1DlgProc(HWND hwnd, UINT Message, WPARAM wParam, LPARAM lParam)
- Dialog2DlgProc(HWND hwnd, UINT Message, WPARAM wParam, LPARAM lParam)
- Dialog3DlgProc(HWND hwnd, UINT Message, WPARAM wParam, LPARAM lParam)
- DlgProc(HWND hwnd, UINT Message, WPARAM wParam, LPARAM lParam)
- DodajZakladke(HWND hwnd, int ident_kontrolki, int pozycja, char[] tytul)
- DoDown(HWND hDialog, HWND hButton, int IDC, float offset)
- DoFileOpen(HWND hwnd, char[] szFileName)
- doFuzzyCalcs()
- DoUp(HWND hDialog, HWND hButton, int IDC, float offset)
- evolutionize(float * popStart, int popStartN, float * result, int resultN, interFunctionListWrapper * iFL1, interFunctionListWrapper * iFL2
- getFuzzyArrays(FuzzyPlantSystem * FPS, interFunctionListWrapper * iFL, int fpsI, float factorOne, float factorTwo)
- getInterFunctions(char * fName, int * fCount)
- getPlantGrowthLogic(char * buf, char * functionName)
- getSNorm(char * fName, int * fCount)
- getTNorm(char * fName, int * fCount)
- getTNormFunctions(char * fName, int * fCount)
- increasingTrapezTest(float x, float * p)
- InitGLS(HWND hwnd)
- mortality(float fitness, int age)
- myAbso(float x)
- MySetPixelFormat(HDC hdc)
- removeFunction(char * functionName, char * type)
- saveBoxes(HWND hwnd, FuzzyRulesSystem * frs)
- triangleTest(float x, float * p)
- updateComboBoxes(HWND hwnd, FuzzyRulesSystem * frs)
- updateEditBoxesDialog1(HWND hwnd, FuzzyRulesSystem * frs)
- WinMain(HINSTANCE hInstance, HINSTANCE hPrevInstance, LPSTR lpCmdLine, int nCmdShow)

List of methods in "ogl.cpp"

- CreateGLWindow(char * title, int width, int height, int bits, bool fullscreenflag)
- createPlant(void * treeParams)
- cross(float x1, float y1, float z1, float x2, float y2, float z2, float * xR, float * yR, float * zR)
- DrawGLScene()
- growPlant(unsigned int age, float * alpha, float * alphaDeviation, float * branchDirection, float * branchDeviation, float markerThreshold, float
- growPlant2(void * cp)
- InitGL()
- initIndeksy()
- KillGLWindow()
- Load_TGA(const char * strfilename)
- LoadBMP(char * Filename)
- LoadGLTextures()
- normalizeOGL(float * x, float * y, float * z)
- readOBJ()
- ReSizeGLScene(GLsizei width, GLsizei height)
- RoX(float alpha, float * v, int a)
- RoY(float alpha, float * v, int a)
- RoZ(float alpha, float * v, int a)
- WndProc(HWND hWnd, UINT uMsg, WPARAM wParam, LPARAM lParam)
- write2Arrays(void * treeParams)
- write2OBJ()

List of functions in DLL

extern "C" __declspec(dllexport) int testBranch (TreeBranch* t, float* prob, int c);

extern "C" __declspec(dllexport) void updateLength(Tree* tr, TreeBranch* t, TreePart* tp, int k);

extern "C" __declspec(dllexport) int markFunction (TreeBranch* t, float* prob, int c);

extern "C" __declspec(dllexport) int /*tNormFunction*/MinimumF (float a, Function *f);

extern "C" __declspec(dllexport) int /*tNormFunction*/AlgebraicF (float a, Function *f);

extern "C" __declspec(dllexport) float /*tNorms*/Minimum (float a, float b);

extern "C" __declspec(dllexport) float /*tNorms*/Algebraic (float a, float b);

extern "C" __declspec(dllexport) void /*sNormFunction*/Maximum (FunctionList *fl, int k, Function *out, float start, float stop);

extern "C" __declspec(dllexport) float S (float x, float *p);

extern "C" __declspec(dllexport) void decreasingTrapezBE (float *r, float *p);

extern "C" __declspec(dllexport) float /*interFunction*/decreasingTrapez (float x, float *p);

extern "C" __declspec(dllexport) void increasingTrapezBE (float *r, float *p);

extern "C" __declspec(dllexport) float /*interFunction*/increasingTrapez (float x, float *p);

extern "C" __declspec(dllexport) void triangleBE (float *r, float *p);

extern "C" __declspec(dllexport) float /*interFunction*/triangle (float x, float *p);

extern "C" __declspec(dllexport) void decQuadSBE (float *r, float *p);

extern "C" __declspec(dllexport) float /*interFunction*/decQuadS (float x, float *p);

extern "C" __declspec(dllexport) int growthFunction1 (TreePart* t, float* a, char id);

extern "C" __declspec(dllexport) int plantGrowthLogic (int a, int b);

extern "C" __declspec(dllexport) int soilLogic (int a, int b);

extern "C" __declspec(dllexport) int healthLogic (int a, int b);

Bibliography

[1] Przemyslaw Prusinkiewicz. Modeling of spatial structure and development of plants. Scientia Horticulturae vol. 74, pp. 113-149, 1998.
[2] Bedrich Benes and Erik Uriel Millan Virtual Climbing Plants Competing for Space. Computer animation, pp. 33-42, 2002.
[3] Radomír Měch, Przemyslaw Prusinkiewicz. Visual models of plants interacting with their environment. SigGraph, 1996.
[4] Reeves, Blau. Approximate and probabilistic algorithms for shading and rendering structured particle systems. SigGraph, 1985.
[5] Prusinkiewicz, Lindenmayer, Hanan. Development models of herbaceous plants for computer imagery purposes. SigGraph, 1988.
[6] De Reffye, Edelin, Françon, Jaeger, Puech. Plant models faithful to botanical structure and development. SigGraph, 1988.
[7]
[8] Prusinkiewicz, Hammel, Mjolsness. Animation of plant development. SigGraph,1993.
[9] Pavol Federl and Przemyslaw Prusinkiewicz. Solving Differential Equations in Developmental Models of Multicellular Structures Expressed Using L−systems. In Proceedings of ICCS 2004, Lecture Notes in Computer Science 3037, pp. 65-72, 2004.
[10] J. Hanan. Parametric L-Systems and their Application to the Modelling and Visualization of Plant models. University of Regina, 1993.
[11] Richard S. Smith1, Cris Kuhlemeier2, Przemyslaw Prusinkiewicz. Inhibition fields for phyllotactic pattern formation: a simulation study. Canadian Journal of Botany 84(11), pp. 1635-1649, 2006.
[12] Prusinkiewicz, James, Měch. Synthetic topiary. SigGraph, 1994.
[13] Joanna L. Power A.J. Bernheim Brush Przemyslaw Prusinkiewiczy David H. Salesin. Interactive Arrangement of Botanical L-System Models. In Proceedings of the 1999 Symposium on Interactive 3D Graphics, pp. 175-182 and 234, 1999.
[14] Lars Mündermann, Yvette Erasmus, Brendan Lane, Enrico Coen, and Przemyslaw Prusinkiewicz. Quantitative Modeling of Arabidopsis Development. Plant Physiology 139, pp. 960-968, 2005.
[15] Callum Galbraith, Przemyslaw Prusinkiewicz, and Brian Wyvill. Modeling a Murex cabritii sea shell with a structured implicit surface modeler. The Visual Computer vol. 18, pp. 70-80, 2002.
[16] Weber, Penn. Creation and rendering of realistic trees. SigGraph, 1994.
[17] Lintermann, Deussen. A Modelling Method and User Interface for Creating Plants. Computer Graphics Forum, Volume 17, Number 1, pp. 73-82, 1998 .
[18] B. Benes. Visual Model of Plant Development with Respect to Influence of Light. Eurographics Workshop on Computer Animation and Simulation. Springer-Verlag, 1997.
[19] N. Greene. Voxel Space Automata: Modeling with Stochastic Growth Processes in Voxel Space. In Proceedings of SIGGRAPH, 1989.
[20] http://www.boga.ruhr-uni-bochum.de/spezbot/skripte/Abb_7_S.html
[21] Cyril Soler , François Sillion , Frédéric Blaise , Philippe de Reffye. A physiological plant growth simulation engine based on accurate radiant energy transfer. Fevrier, 2001.
[22] Ned Greene. Voxel Space Automata: modeling with stochastic growth processes in voxel space. Computer Graphics, Vol.23, No.2, pp.175-184 ,1989.

[23] Mitch Allen, Przemyslaw Prusinkiewicz, and Theodore DeJong. Using L–systems for modeling source–sink interactions, architecture and physiology of growing trees: the L–PEACH model. New Phytologist 166, pp. 869-880, 2005.

[24] Julia Taylor-Hell. Biomechanics in Botanical Trees. M.Sc. thesis, University of Calgary, 2005.

[25] Sylvain Lefebvre, Fabrice Neyret. Synthesizing Bark. Eurographics: Workshop on Rendering, 2002.

[26] E. H. Mamdani. Application of fuzzy algorithms for control of simple dynamic plant. Academic Press, New York, 1974.

[27] Takagi, T., Sugeno, M. Fuzzy Identification of Systems and Its Applications to Modeling and Control. IEEE Trans. on Syst., Man and Cyb, 15:116-132, 1985.

[28] Agner Fog. Optimizing Software in C++. Copenhagen University College of Engineering, 2007.

[29] Zbigniew Michalewicz. Genetic Algorithms + Data Structures = Evolution Programs. Springer-Verlag, 1st Ed 1992.

[30] Lawrence J. Fogel. Autonomous Automata. Industrial Research,. 4:14-19, 1962.

[31] Holland, J. H. Information processing in adaptive systems. Volume 3 of the Proceedings of the International Union of Physiological Sciences, p. 330-338, 1962.

[32] Ingo Rechenberg. Cybernetic Solution Path of an Experimental Problem. 1965.

[33] O. Cordon, F. Gomide, F. Herrera, F. Hoffmann and L. Magdalena. Ten years of genetic fuzzy systems: current framework and new trends. Fuzzy Sets and Systems, Elsevier, 141, 2004.

[34] K. Shinozaki, K. Yoda, K. Hozumi ard T. Kira. A quantitative analysis of plant form-the pipe model theory. Japanese Journal of Ecology 14:97-104, 1964.

[35] W. Bugała. Drzewa i Krzewy dla terenów zieleni. Państwowe Wydawnictwo Rolnicze i Leśne, Warszawa 1991

[36] C. Murray. A relationship between circumference and weight in trees and its bearing on branching angles. *Journal of General Physiology*, 10:725–729, 1927.

[37] Radoslaw Karwowski and Przemyslaw Prusinkiewicz. The L-system-based plant-modeling environment L-studion 4.0. In Proceedings of the 4th International Workshop on Functional-Structural Plant Models, pp. 403-405, 2004.

www.ingramcontent.com/pod-product-compliance
Lightning Source LLC
LaVergne TN
LVHW080105070326
832902LV00014B/2425